FRACTION WORD PROBLEMS

LEVEL 1

BILL S. LEE

Introduction

This book is the sequel of Word Problems-Detailed Explanations of Reasoning and Solving Strategies (Volumes 11-A, 11-B and 12). The 50 problems in the book focus on only the techniques on fiction word problems solving. If you are not sure of certain terms or concepts in the book, please go to Amazon.com to get these books:

Word Problems-Detailed Explanations of Reasoning and Solving Strategies by Bill S. Lee (Volumes 11-A, 11-B and 12) by Bill S. Lee

You can also go to www.amathproblem.com to order the books, to see more math questions and to learn more techniques on math problems solving.

QUESTIONS

1. If $\dfrac{5}{12}$ ton of goods are shipped out of a warehouse on each day, how many tons will be shipped out in 3 days?

2. Hanna bought a book with 240 pages. She has read $\dfrac{3}{4}$ of the entire book so how many pages has she read?

3. The distance between Peter's house and John's house is 180 yards.

Peter is walking to John's house and has already covered $\frac{5}{6}$ of the entire distance so how many more yards does he need to walk before he gets to John's house?

4. If one cup of water weighs $\frac{3}{4}$ pound, how much does $\frac{2}{5}$ of a cup of water weigh?

5. Hanna has $120.00 in her savings account; John has $\frac{5}{8}$ as much as

Hanna in his savings account; Leon has $\frac{5}{3}$ times as much as John. How much does Leon have in his savings account?

6. A rope is $5\frac{5}{6}$ feet long. If $\frac{1}{5}$ of is cut off first and $\frac{1}{5}$ foot is cut off after that, how many feet of the rope will be cut off altogether?

7. A high school had 250 graduates last year. If the number of graduates this year is $\frac{1}{5}$ more than last year, how many students will graduate this year?

8. 6th grade in Westview Elementary School has 216 students. If the number of students in 5th grade is $\frac{7}{9}$ as many as the number of students in 6th grade, how many students are in 5th grade?

9. It used to take a train 12 minutes to pass a canal but now the amount of time the train uses is $\frac{1}{3}$ less than the amount of time it used to use so how much time does it take the train to pass the canal?

10. A road is 18,000 miles long so $\frac{3}{20}$ of it is how many miles?

11. An orchard has a total of 500 trees. If $\frac{3}{5}$ of them are apple trees, how many apple trees are in the orchard?

12. The price of one book is $15.00. If the price of another book is $\frac{2}{3}$ as much, what is the price of the other book?

13. Among 180 students, $\frac{1}{6}$ of them joined a math club; $\frac{1}{9}$ of them joined a chess club. If the rest of the students joined the club of arts,

(1). How many students joined the chess club?

(2). How many students joined the math club?

(3). How many students joined the club of arts?

14. The distance between John's house and Peter's house is 120 yards. John is walking to Peter's house and has already walked $\frac{5}{8}$ of the distance so how far is he from Peter's house?

15. Mom used $\frac{1}{9}$ of $\frac{9}{10}$ pounds of oil in cooking so how many pounds of oil

did she use in cooking?

16. Granddad is 60 years old this year and Dad is $\frac{7}{12}$ as old as Granddad. If Leon is $\frac{2}{7}$ as old as Dad, how old is Leon?

17. Mom bought 12 pounds of fruits for the week and apples weighed $\frac{3}{4}$ of the total weight of all the fruits. If the apples that the family ate during the week weighed $\frac{2}{3}$ of all the apples, how many pounds of apples did the family eat?

18. Matthew wrote 60 diaries this semester and Hanna wrote $\frac{3}{4}$ as many as Matthew. If the number of diaries that Lily wrote was $\frac{5}{9}$ as many as the number of diaries that Hanna wrote, how many did Lily write?

19. A warehouse had 360 tons of fertilizer stored in it last week. $\frac{4}{9}$ of it was shipped out on this Monday. If the fertilizer shipped out on Tuesday weighed $1\frac{1}{4}$ times as much as the fertilizer shipped out on Monday, how many tons of fertilizer was shipped out on Tuesday?

20. Lily was driving from her house at 45 miles per hour to go to her

friend's house.   If she got there in $1\frac{2}{3}$ hours, how many miles were between her house and her friend's?

21. Peter and John were driving to meet each other on the way. Peter was going at $32\frac{2}{5}$ miles per hour; John was going at $34\frac{1}{2}$ miles per hour. If they met on the way in $3\frac{1}{3}$ hours after they left the houses, what was the distance between the houses?

22. It used to take a train 12 minutes to pass a tunnel but now the amount of time the train needs to pass the tunnel is $\frac{1}{3}$ less than it used to. How many fewer minutes does the train need to pass the tunnel now than it used to?

23. A toy factory planned to make 5,000 toys this year but actually made $\frac{1}{5}$ more than what it planned. How many toys did the factory actually make this year?

24. Three groups of city workers planted trees. The first group planted 39 trees; the second group planted $\frac{2}{3}$ as many as the first group. If the number of trees the third group planted was 5 less than twice as many as the second group, how many trees did the third group of workers plant?

25. 128 students took a math test and $\frac{5}{8}$ of them passed the test. If $\frac{2}{5}$ of those that passed the test were girl students, the number of boys

that passed the test was what percent of all the students that took the test?

26. Lily read $\frac{1}{5}$ of a book of 180 pages on Monday and $\frac{2}{9}$ of the entire book on Tuesday. How many more pages did she read on Tuesday than on Monday?

27. Farmers used $\frac{3}{4}$ of $\frac{4}{5}$ ton of fertilizer during the first week. If the amount of fertilizer the farmers used during the second week was $\frac{1}{10}$ as much as what they used during the first week, how many tons of fertilizer was left after two weeks?

28. John made 300 spare parts; Peter made $\frac{3}{4}$ as many as John. If Leon made $\frac{8}{9}$ as many as Peter, how many did Leon make?

29. Hanna has 25 anima stamps; John has 30. If the number of animal stamps that Sam has is $\frac{3}{5}$ as many as the total number of animal stamps that Hanna and John both have, how many stamps does Sam have?

30. A person walks $2\frac{4}{5}$ miles per hour; a bus travels 20 times as much as the person per hour. If a car covers $1\frac{1}{2}$ as much distance as the bus per hour, how many miles does the car travel per hour?

31. Wanda is reading a book of 320 pages. If she read $\frac{1}{8}$ of the book on the first day and $\frac{1}{2}$ as much on the second day as what she read on the first day, how many pages did she read in both days?

32. A clothing factory planned to make 720 children clothes in March but actually made $\frac{3}{5}$ of the amount during the first half of the month and the same amount in the second half of the month as in the first half of the month. How many more clothes did the factory actually make in March?

33. One class has 55 students in it; $\frac{1}{5}$ of them are girl students. When 5 boys are transferred to other classes, the number of girls is what percentage of the students in the class?

34. When the length and the width of a rectangle are both shortened to $\frac{2}{3}$ of the original size, the area of new rectangle is what fractional part of the area of the original rectangle?

35. The base of a parallelogram is increased to $\frac{7}{2}$ as long as the original base; at the same time, the height on the base is reduced to $\frac{2}{7}$ as long as the original height. The area of the new parallelogram is what percentage of the area of the original parallelogram?

36. A factory planned to make 480 washing machines last week and $\frac{1}{6}$ more this week. If number of washing machines the factory plans to make next week is $\frac{1}{8}$ more than this week, how many washing machines does the

factory plan to make next week?

37. There were 96 students in a math club and the number of boy students was $\frac{3}{8}$ of the number of all the students in the club. Recently, more boy students have joined the club so that the number of boys is $\frac{5}{6}$ of the number of girls in the club now. How many boy students have joined the club recently?

38. Three groups of city workers are planting trees. The second group has already planted 480 trees; the third group planted $1\frac{1}{6}$ as many as

the second group; the first group planted $\frac{5}{6}$ as many as the second group. How many trees have all the workers planted?

39. A rope is 12 feet long. If $\frac{1}{5}$ of it is cut off at first and $\frac{1}{3}$ of it is cut off after that, how much shorter is the first cut than the second one?

40. There are 3 piles of coal: the first pile has 240 tons of coal in it; the second pile of coal weighs $\frac{3}{4}$ as much as the first pile; the

second pile of coal weighs $\frac{4}{5}$ ton less than the third pile of coal. How much do all the three piles of coal weigh?

41. Hanna is reading a book of 320 pages and has read $\frac{2}{5}$ of the book on the first day. If she read $\frac{3}{8}$ of the remaining pages of the book on the second day, how many pages hasn' t she read yet?

42. A pile of coal weighs 40 tons; a truck shipped $\frac{2}{5}$ of it away on the first trip. Based on the information, answer the following 3 questions:

(1). If the truck ships away $\frac{3}{4}$ ton on the second trip, how many tons will be left after that?

(2). If the truck ships away $\frac{3}{4}$ of the rest of the coal on the second trip, how many tons of coal will be left after that?

(3). If the truck ships away $\frac{3}{4}$ as much as what it shipped away on the first trip, how many tons of coal will be left after that?

43. A bookstore sold 1,400 books on the morning of a Saturday and $\frac{6}{7}$ as many on the afternoon of the Saturday. If the number of books sold on Sunday was $\frac{9}{13}$ as many as the number of books sold on Saturday, how many books were sold on Sunday?

44. The amount of money Peter has is $\frac{1}{5}$ more than John; the amount of money John has is $\frac{1}{5}$ more than Leon. The amount of money that Peter has is what fractional part more than the amount of money that Leon has?

45. A book has 135 pages. Lily has read $\frac{2}{5}$ of the book during the first 5 days after she bought the book so which page will she need to start with on the 6th day?

46. A farmer harvested 6,400 pounds of apples last year. If this year

the harvest was $\frac{1}{5}$ less than last year, how many fewer pounds of apples did the farmer harvest this year than last year?

47. A store got a total of 960 pounds of apples and pears to sell. If the apples weighed $\frac{3}{5}$ of the fruits, how many more pounds of apples did the store get than pears?

48. A construction crew is maintaining a road and has already worked on it for 5 days. If, on the average, the crew fixes $\frac{1}{20}$ of the road on

each day, what fractional part of the road hasn't the crew maintained yet?

49. The distance between City A and City B is 150 miles. Peter started from City A and is driving to City B. If he has travelled $\frac{2}{5}$ of the entire distance, how far is he away from City B?

50. A rope is $3\frac{2}{5}$ feet long. Peter cut $\frac{1}{2}$ of it at first then cut $\frac{1}{2}$ foot from the rest of the rope. How many feet of the rope are still left?

ANSWER KEYS

Introduction

This book is the sequel of Word Problems-Detailed Explanations of Reasoning and Solving Strategies (Volumes 11-A, 11-B and 12). The problems in the book focus on fiction word problem solving techniques. If you are not sure of certain terms and concepts used in the book, please go to Amazon.com to get these books:

Word Problems-Detailed Explanations of Reasoning and Solving Strategies by Bill S. Lee (Volumes 11-A, 11-B and 12)

You can also go to www.amathproblem.com to order the books, to see more math questions and to learn more techniques on math problems solving.

Answer keys:

1. If $\frac{5}{12}$ ton of goods are shipped out of a warehouse on each day, how many tons will be shipped out in 3 days?

Reasoning and solving strategies:

On each day, the same amount is shipped out so we use the 3 elements formulas to solve the problem:

The amount in each group is the amount shipped out on each day: $\frac{5}{12}$ ton

The number of groups is the number of days for the goods to be shipped out: 3 days

The total amount in all the groups is the total amount shipped out in 3 days:

The total amount in all the groups= the amount in each group x the number of groups= $\frac{5}{12}$ ton x $3 = \frac{15}{12}$ tons

2. Hanna bought a book with 240 pages. She has read $\frac{3}{4}$ of the entire book so how many pages has she read?

Reasoning and solution:

From "read $\frac{3}{4}$ of the entire book", we know that the amount in 1 in the entire book, 240 pages.

The number of pages she has read is the amount in the fractional part is $\frac{3}{4}$

The amount in 1 x the fractional part= 240 x $\frac{3}{4}$ =180 pages

Hanna has read 180 pages.

3. The distance between Peter's house and John's house is 180 yards. Peter is walking to John's house and has already covered $\frac{5}{6}$ of the entire distance so how many more yards does he need to walk before he gets to John's house?

Reasoning and solution:

He has covered $\frac{5}{6}$ of the entire distance so the entire distance is the amount in 1. The fractional part of the entire distance that is not covered is $(1-\frac{5}{6})$, or is $\frac{1}{6}$ of the entire distance. The distance he has not covered yet is the amount in the fractional part, $\frac{1}{6}$, so we multiply the fractional part, $\frac{1}{6}$, with the amount in 1, the entire distance, 180, to find the amount in the fractional part, the distance he has not covered yet:

180 x $\frac{1}{6}$ =30 yards

4. If one cup of water weighs $\frac{3}{4}$ pound, how much does $\frac{2}{5}$ of a cup of

26

water weigh?

Reasoning and solution:

From "$\frac{2}{5}$ of a cup of water" we know that the weight of one cup of water is the amount in 1; we need to find the weight of water in the fractional part, $\frac{2}{5}$, so we multiply the amount in 1, $\frac{3}{4}$ pound, with the fractional part, $\frac{2}{5}$, to find the amount in the fractional part:

$$\frac{3}{4} \text{ pound} \times \frac{2}{5} = \frac{3}{10} \text{ pound}$$

5. Hanna has $120.00 in her savings account; John has $\frac{5}{8}$ as much as Hanna in his savings account; Leon has $\frac{5}{3}$ times as much as John. How much does Leon have in his savings account?

Reasoning and solution:

We need to first find how much John has in his savings account:

"John has $\frac{5}{8}$ as much as Hanna" so the amount in 1 is the amount that Hanna has, $120.00 and the amount that John has is the amount in the fractional part, $\frac{5}{8}$. We multiply the amount in 1, $120.00, with the fractional part, $\frac{5}{8}$, to find the amount in the fractional part, the amount of money John has in his savings account:

$$120 \times \frac{5}{8} = \$75.00$$

Now we are ready to find the amount of money that Leon has in his savings account:

"Leon has $\frac{5}{3}$ times as much as John" so the amount that John has is the amount in each group and the amount that Leon has is the amount in all the $\frac{5}{3}$ groups. We multiply the amount in each group, $75.00, with the number of groups, $\frac{5}{3}$, to find the amount in the groups:

$$75 \times \frac{5}{3} = \$125.00$$

Leon has $125.00 in his savings account.

6. A rope is $5\frac{5}{6}$ feet long. If $\frac{1}{5}$ of is cut off first and $\frac{1}{5}$ foot is cut off after that, how many feet of the rope will be cut off altogether?

Reasoning and solution:

First find the amount of the rope that is cut off at first:

"$\frac{1}{5}$ of is cut off first" so the length of the entire rope is the amount in 1, and the amount that is cut off at first is the amount in the fractional part, $\frac{1}{5}$. We multiply the amount in 1, $5\frac{5}{6}$ feet, with the fractional part, $\frac{1}{5}$, to find the amount in the fractional part:

$$5\frac{5}{6} \times \frac{1}{5} = \frac{7}{6} \text{ feet}$$

Secondly, find the total amount that is cut off: Combine the amount that is cut off at first, $\frac{7}{6}$ feet, with the amount that is cut off secondly, $\frac{1}{5}$ feet:

$$\frac{1}{5} + \frac{7}{6} = \frac{41}{30} \text{ feet}$$

$\frac{41}{30}$ feet of the rope are cut off altogether.

7. A high school had 250 graduates last year. If the number of graduates this year is $\frac{1}{5}$ more than last year, how many students will graduate this year?

Reasoning and solution:

From "the number of graduates this year is $\frac{1}{5}$ more than last year" we know that the number of graduates last year is the amount in 1 and that the number of graduates this year is the amount in the fractional part that is $\frac{1}{5}$ more than 1. "$\frac{1}{5}$ more than 1" is $(1+\frac{1}{5})$ which is $\frac{6}{5}$ so the number of graduates this year is the amount in the fractional part $\frac{6}{5}$.

We multiply the amount in 1, 250, with the fractional part, $\frac{6}{5}$, to find the amount in the fractional part, the number of graduates this year:

$250 \times \dfrac{6}{5} = 300$ graduates this year

8. 6th grade in Westview Elementary School has 216 students. If the number of students in 5th grade is $\dfrac{7}{9}$ as many as the number of students in 6th grade, how many students are in 5th grade?

Reasoning and solution:

From "If the number of students in 5th grade is $\dfrac{7}{9}$ as many as the number of students in 6th grade" we know that the number of students in 6th grade is the amount in 1 and that the number of students in 5th grade is the amount in the fractional part, $\dfrac{7}{9}$, so we multiply the fractional part with the amount in 1 to find the amount in the fractional part: the number of students in 5th grade:

$216 \times \dfrac{7}{9} = 168$ students

9. It used to take a train 12 minutes to pass a canal but now the amount of time the train uses is $\dfrac{1}{3}$ less than the amount of time it used to use so how much time does it take the train to pass the canal?

Reasoning and solution:

From "the amount of time the train uses is $\dfrac{1}{3}$ less than the amount of time it used to use" we know that the amount of time it used to use is the amount in 1 and that the amount of time the train uses now is the amount in the fractional part that is $\frac{1}{3}$ less than 1: "$\frac{1}{3}$ less than 1"

means $1-\frac{1}{3}=\frac{2}{3}$, so the amount of time the train uses now is the amount

in the fractional part, $\frac{2}{3}$. We multiply the amount in 1, with the

fractional part, to find the amount in the fractional part, the amount
of time the train uses now to pass a canal:

$12 \times \frac{2}{3} = 8$ minutes

10. A road is 18,000 miles long so $\frac{3}{20}$ of it is how many miles?

Reasoning and solution:

From "$\frac{3}{20}$ of it" we know that the entire road, 18,000 miles, is the

amount in 1 and that the fractional part is $\frac{3}{20}$. We multiply the

fractional part with the amount in 1, to find the amount in the
fractional part:

$18,000 \times \frac{3}{20} = 2,700$ miles

$\frac{3}{20}$ of the road is 2,700 miles long.

11. An orchard has a total of 500 trees. If $\frac{3}{5}$ of them are apple trees,

how many apple trees are in the orchard?

Reasoning and solution:

From "$\frac{3}{5}$ of them are apple trees" we know that the "them" or the " 500 trees" is the amount in 1; the number of apple trees is the amount in the fractional part, $\frac{3}{5}$, so we multiply the fractional part with the amount in 1 to find the amount in the fractional part, the number of apple trees in the orchard:

$$500 \times \frac{3}{5} = 300$$

The orchard has 300 apple trees.

12. The price of one book is \$15.00. If the price of another book is $\frac{2}{3}$ as much, what is the price of the other book?

Reasoning and solution;

From "the price of another book is $\frac{2}{3}$ as much (as the first book)" we know that the price of the first book, \$15.00, is the amount in 1 and that the price of the second book is the amount in the fractional part, $\frac{2}{3}$, so we multiply the fractional part with the amount in 1 to find the amount in the fractional part, the price of the second book:

$$15 \times \frac{2}{3} = \$10.00$$

13. Among 180 students, $\frac{1}{6}$ of them joined a math club; $\frac{1}{9}$ of them joined a chess club. If the rest of the students joined the club of arts,

(1). How many students joined the chess club?

Reasoning and solution:

From "$\frac{1}{9}$ of them joined a chess club" we know that "them" or "180 students" is the amount in 1 and that the number of students that joined the chess club is the amount in the fractional part, $\frac{1}{9}$, so we multiply the amount in 1 with the fractional part to find the amount in the fractional part, the number of students that joined the chess club:

$180 \times \frac{1}{9} = 20$ students

(2). How many students joined the math club?

Reasoning and solution:

From "$\frac{1}{6}$ of them joined a math club" we know that "them" or "180 students" is the amount in 1 and that the number of students that joined the math club is the amount in the fractional part, $\frac{1}{6}$, so we multiply the amount in 1 with the fractional part to find the amount in the fractional part, the number of students that joined the math club:

$180 \times \frac{1}{6} = 30$ students

(3). How many students joined the club of arts?

Reasoning and solution:

Subtract the number of students that joined the math club and the chess club from 180 to find the number of students that joined the club of arts:

180−20−30=130

14. The distance between John's house and Peter's house is 120 yards. John is walking to Peter's house and has already walked $\frac{5}{8}$ of the distance so how far is he from Peter's house?

Reasoning and solution:

From "has already walked $\frac{5}{8}$ of the distance" we know that the total distance is the amount in 1. Since he has already covered $\frac{5}{8}$ of the distance, the fractional part of the distance that he has not covered is:

$1-\frac{5}{8}=\frac{3}{8}$

We multiply the fractional part, $\frac{3}{8}$, with the amount in 1, the total distance, to find the amount in the fractional part, the distance that John is away from Peter's house:

$120 \times \frac{3}{8}$ =45 yards

15. Mom used $\frac{1}{9}$ of $\frac{9}{10}$ pounds of oil in cooking so how many pounds of oil did she use in cooking?

Reasoning and solution:

From "$\frac{1}{9}$ of $\frac{9}{10}$ pounds of oil in cooking" we know that $\frac{9}{10}$ pounds of oil is the amount in 1 and the amount of oil that Mom used is the amount in the fractional part, $\frac{1}{9}$, so we multiply the amount in 1 with the fractional part to find the amount in the fractional part, the amount of oil mom used in cooking:

$\frac{1}{9} \times \frac{9}{10} = \frac{1}{10}$ pound

16. Granddad is 60 years old this year and Dad is $\frac{7}{12}$ as old as Granddad. If Leon is $\frac{2}{7}$ as old as Dad, how old is Leon?

Reasoning and solution:

We need two sets of the 3 elements to solve the problem:

First set of the 3 elements: find the age of Dad:

From "Dad is $\frac{7}{12}$ as old as Granddad" we know that the age of Granddad is the amount in 1 and that the age of Dad is the amount in the fractional part, $\frac{7}{12}$, so we multiply the amount in 1 with the fractional part to find the amount in the fractional part, the age of Dad:

$60 \times \dfrac{7}{12} = 35$ years old

Second set of the 3 elements: find the age of Leon:

From "Leon is $\dfrac{2}{7}$ as old as Dad" we know that the age of Dad is the amount in 1 and that the age of Leon is the amount in the fractional part, $\dfrac{2}{7}$, so we multiply the amount in 1 with the fractional part, to find the amount in the fractional part, the age of Leon:

$35 \times \dfrac{2}{7} = 10$ years old

17. Mom bought 12 pounds of fruits for the week and apples weighed $\dfrac{3}{4}$ of the total weight of all the fruits. If the apples that the family ate during the week weighed $\dfrac{2}{3}$ of all the apples, how many pounds of apples did the family eat?

Reasoning and solution:

We need two set of the 3 elements to solve the problem:

First set of the 3 elements: find how many pounds of apples Mom bought:

From "apples weighed $\dfrac{3}{4}$ of the total weight of all the fruits" we know that the total weight of all the fruits is the amount in 1 and that the weight of the apples Mom bought is the amount in the fractional part $\dfrac{3}{4}$ so we multiply the fractional part with the amount in 1 to find the amount in the fractional part, the weight of apples Mom bought:

$12 \times \dfrac{3}{4} = 9$ pounds

Second set of the 3 elements:

From "the apples that the family ate during the week weighed $\dfrac{2}{3}$ of all the apples" we know that the total weight of all the apples is the amount in 1 and that the weight of the apples the family ate during the week is the amount in the fractional part $\dfrac{2}{3}$ so we multiply the fractional part with the amount in 1 to find the amount in the fractional part, the weight of apples the family ate during the week:

$9 \times \dfrac{2}{3} = 6$ pounds

18. Matthew wrote 60 diaries this semester and Hanna wrote $\dfrac{3}{4}$ as many as Matthew. If the number of diaries that Lily wrote was $\dfrac{5}{9}$ as many as the number of diaries that Hanna wrote, how many did Lily write?

Reasoning and solution:

We need two sets of 3 elements to solve the problem:

The first set is to find the number of diaries that Hanna wrote:

From "Hanna wrote $\dfrac{3}{4}$ as many as Matthew" we know that the number of diaries that Matthew wrote is the amount in 1 and that the number of diaries Hanna wrote is the amount in the fractional part, $\dfrac{3}{4}$, so we multiply the fractional part with the amount in 1 to find the amount in the fractional part, the number of diaries that Hanna wrote:

$60 \times \dfrac{3}{4} = 45$ diaries

The second set is to find the number of diaries that Lily wrote:

From "the number of diaries that Lily wrote was $\dfrac{5}{9}$ as many as the number of diaries that Hanna wrote" we know that the number of diaries that Hanna wrote is the amount in 1 and that the number of diaries Lily wrote is the amount in the fractional part, $\dfrac{5}{9}$, so we multiply the fractional part with the amount in 1 to find the amount in the fractional part, the number of diaries that Lily wrote:

$45 \times \dfrac{5}{9} = 25$ diaries

19. A warehouse had 360 tons of fertilizer stored in it last week. $\dfrac{4}{9}$ of it was shipped out on this Monday. If the fertilizer shipped out on Tuesday weighed $1\dfrac{1}{4}$ times as much as the fertilizer shipped out on Monday, how many tons of fertilizer was shipped out on Tuesday?

Reasoning and solution:

We need 2 sets of 3 elements to solve the problem.

The first set is to find the amount shipped out on Monday:

From "$\dfrac{4}{9}$ of it was shipped out on Monday." we know that the "it" or "the 360 tons of fertilizer" is the amount in 1 and that the amount in the fractional part, $\frac{4}{9}$, is the amount shipped out on Monday. We multiply the amount in 1 with the fractional part to find the amount in the fractional part, the amount of fertilizer shipped out on Monday:

$360 \times \frac{4}{9} = 160$ tons

The second set is to find the amount shipped out on Tuesday:

From "the fertilizer shipped out on Tuesday weighed $1\frac{1}{4}$ times as much as the fertilizer shipped out on Monday" we know that the amount of fertilizer shipped on Monday is the amount in 1 group and that the amount shipped on Tuesday is the amount in the $1\frac{1}{4}$ groups so we multiply the amount in 1 group with the number of groups to find the amount in the groups, the amount shipped out on Tuesday:

$160 \times 1\frac{1}{4} = 200$ tons

20. Lily was driving from her house at 45 miles per hour to go to her friend's house. If she got there in $1\frac{2}{3}$ hours, how many miles were between her house and her friend's?

Reasoning and solution:

Use the formula: total distance= the speed $\times$ the amount of time spent on the distance:

$45 \times 1\frac{2}{3} = 75$ miles

21. Peter and John were driving to meet each other on the way. Peter was going at $32\frac{2}{5}$ miles per hour; John was going at $34\frac{1}{2}$ miles per hour. If they met on the way in $3\frac{1}{3}$ hours after they left the houses, what was the distance between the houses?

Reasoning and solution:

The total distance covered per hour is the sum of the speed of Peter and the sum of the speed of John:

$$32\frac{2}{5}+34\frac{1}{2}=66\frac{9}{10} \text{ miles per hour}$$

Use the formula: total distance= the speed $\times$ the amount of time spent on the distance:

$$66\frac{9}{10} \text{ miles per hour} \times 3\frac{1}{3}=223 \text{ miles}$$

22. It used to take a train 12 minutes to pass a tunnel but now the amount of time the train needs to pass the tunnel is $\frac{1}{3}$ less than it used to. How many fewer minutes does the train need to pass the tunnel now than it used to?

Reasoning and solution:

From "the amount of time the train needs to pass the tunnel is $\frac{1}{3}$ less than it used to" we know that the amount of time it used to take is in 1 and the amount of time it takes now is in the fractional part: $1-\frac{1}{3}$

$$=\frac{2}{3}$$

We multiply the amount in 1, 12 minutes, with the fractional part, $\frac{2}{3}$, to find the amount in the fractional part, the amount of time it takes for the train to pass the tunnel:

$12 \times \frac{2}{3} = 8$ minutes

$12-8=4$ minutes

Now the train needs 4 minutes less to pass the tunnel.

23. A toy factory planned to make 5,000 toys this year but actually made $\frac{1}{5}$ more than what it planned. How many toys did the factory actually make this year?

Reasoning and solution:

From "actually made $\frac{1}{5}$ more than what it planned." The number of toys the factory planned to make is the amount in 1; the number of toys the factory actually made is the amount in the fractional part that is $\frac{1}{5}$ than 1 : $1+\frac{1}{5}=\frac{6}{5}$

We multiply the amount in 1, 5,000, with the fractional part, to find the amount in the fractional part, the number of toys the factory actually made this year:

$5,000 \times \dfrac{6}{5} = 6,000$ toys

24. Three groups of city workers planted trees. The first group planted 39 trees; the second group planted $\dfrac{2}{3}$ as many as the first group. If the number of trees the third group planted was 5 less than twice as many as the second group, how many trees did the third group of workers plant?

Reasoning and solution:

The problem needs 2 sets of 3 elements to solve:

The first set of 3 elements:

From "the second group planted $\dfrac{2}{3}$ as many as the first group" we know the number of trees the first group planed is the amount in 1; the number of trees the second group planted is the amount in the fractional part, $\dfrac{2}{3}$, so we multiply the amount in 1 with the fractional part, to find the amount in it:

$39 \times \dfrac{2}{3} = 26$ trees

The second set of 3 elements:

From "the number of trees the third group planted was 5 less than twice as many as the second group" we know that the number of trees the second group planted is the amount in each group and that the number of trees the third group planted is 5 less than the amount in 2 groups. We multiply the amount in each group with the number of groups to find the amount in two groups:

26×2=52

5 less than the amount in 2 groups are:

52-5=47 trees

The third group planted 47 trees.

25. 128 students took a math test and $\frac{5}{8}$ of them passed the test. If $\frac{2}{5}$ of those that passed the test were girl students, the number of boys that passed the test was what percent of all the students that took the test?

Reasoning and solution:

We need 3 sets of 3 elements to solve the problem:

First set of 3 elements:

From "$\frac{5}{8}$ of them passed the test" we know that amount in 1 is 128 students. We multiply the amount in 1, with the fractional part, $\frac{5}{8}$, to find the amount in the fractional part, which is the number of students that passed the test:

$128 \times \frac{5}{8}$ =80 students

Second set of 3 elements:

From "$\frac{2}{5}$ of those that passed the test were girl students" we know that the number of students that passed the test, 80 students, is the amount in 1.

Since $\frac{2}{5}$ of the students that passed the test were girls, $(1-\frac{2}{5})=\frac{3}{5}$ of the students that passed the test were boys. We multiply the amount in 1 with the fractional part, $\frac{3}{5}$, to find the amount in the fractional part, the number of boys that passed the test:

$80\times\frac{3}{5}=48$ boys

Third set of the 3 elements:

From "the number of boys that passed the test was what percent of all the students that took the test" we know that the amount in 1 is the total number of students that took the test; the amount in the fractional part is the number of boys that passed the test. We divide the amount in 1 into the amount in the fractional part to find the fractional part:

$48\div128=37.5\%$

The number of boys that passed the test was 37.5% of all the students that took the test.

26. Lily read $\frac{1}{5}$ of a book of 180 pages on Monday and $\frac{2}{9}$ of the entire book on Tuesday. How many more pages did she read on Tuesday than on Monday?

Reasoning and solution:

We need 2 sets of 3 elements to solve the problem.

First set of 3 elements:

44

From "Lily read $\frac{1}{5}$ of a book of 180 pages" we know that the amount in 1 is 180 pages and the number of pages she read is the amount in the fractional part is $\frac{1}{5}$. We multiply the fractional part with the amount in 1 to find the amount in the fractional part, the number of pages Lily read on Monday:

$180 \times \frac{1}{5} = 36$ pages

The second set of 3 elements:

From "$\frac{2}{9}$ of the entire book on Tuesday" we know that the entire book, 180 pages, is the amount in 1; the number of pages she read on Tuesday is the amount in the fractional part, is $\frac{2}{9}$. We multiply the amount in 1 with the fractional part to find the amount in the fractional part:

$180 \times \frac{2}{9} = 40$ pages

Lily read 40 pages on Tuesday.

40-36=4 pages

She read 4 more pages on Tuesday than on Monday.

27. Farmers used $\frac{3}{4}$ of $\frac{4}{5}$ ton of fertilizer during the first week. If the amount of fertilizer the farmers used during the second week was $\frac{1}{10}$ as much as what they used during the first week, how many tons of fertilizer was left after two weeks?

Reasoning and solution:

We need two set of the 3 elements to solve the problem.

First set of 3 elements:

From "Farmers used $\frac{3}{4}$ of $\frac{4}{5}$ ton of fertilizer during the first week" we know that the amount in 1 is $\frac{4}{5}$ ton of fertilizer and that the fractional part is $\frac{3}{4}$ so we multiply the amount in 1 with the fractional part to find the amount in the fractional part, the amount of fertilizer used during the first week:

$$\frac{4}{5} \times \frac{3}{4} = \frac{3}{5} \text{ tons}$$

Second set of 3 elements:

From "the amount of fertilizer the farmers used during the second week was $\frac{1}{10}$ as much as what they used during the first week" we know that the amount in 1 is $\frac{3}{5}$ tons and the fractional part is $\frac{1}{10}$, so we multiply the fractional part with the amount in 1, to find the amount in the fractional part, the amount of fertilizer used in the second week:

$$\frac{3}{5} \times \frac{1}{10} = \frac{3}{50} \text{ tons}$$

The amount of fertilizer left after two weeks:

$$\frac{4}{5} - \frac{3}{5} - \frac{3}{50} = \frac{7}{50} \text{ tons}$$

28. John made 300 spare parts; Peter made $\frac{3}{4}$ as many as John. If Leon made $\frac{8}{9}$ as many as Peter, how many did Leon make?

Reasoning and solution:

We need 2 sets of 3 elements:

First set of 3 elements:

From "Peter made $\frac{3}{4}$ as many as John" the number of spare parts made by John is the amount in 1; the number of spare parts made by Peter is the amount in the fractional part, $\frac{3}{4}$, so we multiply the amount in 1 with the fractional part to find the amount in the fractional part:

$300 \times \frac{3}{4} = 225$

Second set of 3 elements:

From "Leon made $\frac{8}{9}$ as many as Peter" the number of spare parts made by Peter is the amount in 1; the number of spare parts made my Leon is the amount in the fractional part, $\frac{8}{9}$, so we multiply the amount in 1 with the fractional part, to find the amount in the fractional part:

$225 \times \frac{8}{9} = 200$ spare parts

Leon made 200 spare parts.

29. Hanna has 25 anima stamps; John has 30. If the number of animal

stamps that Sam has is $\frac{3}{5}$ as many as the total number of animal stamps that Hanna and John both have, how many stamps does Sam have?

Reasoning and solution:

The total number of stamps that Hanna and John both have is:

25+30=55

We need 1 set of 3 elements to solve the problem:

From "the number of animal stamps that Sam has is $\frac{3}{5}$ as many as the total number of animal stamps that Hanna and John both have" we know that amount in 1 is the total number of stamps that Hanna and John have, 55, and that the number of stamps that Sam has is the amount in the fractional part, $\frac{3}{5}$, so we multiple the fractional part with the amount in 1 to find the amount in it, the number of stamps that Sam has:

$55 \times \frac{3}{5}$ =33

30. A person walks $2\frac{4}{5}$ miles per hour; a bus travels 20 times as much as the person per hour. If a car covers $1\frac{1}{2}$ as much distance as the bus per hour, how many miles does the car travel per hour?

Reasoning and solution:

We need 2 sets of 3 elements to solve the problem:

The first set of the 3 elements:

From "a bus travels 20 times as much as the person per hour" the distance covered by the person per hour is the amount in each group; the distance covered by the bus per hour is the amount in 20 groups so we multiple the number of groups with the amount in each group to find the total amount in all the groups:

$2\frac{4}{5} \times 20 = 56$ miles per hour

The second set of the 3 elements:

From "a car covers $1\frac{1}{2}$ as much distance as the bus per hour" the amount in 1 is the distance covered by the bus per hour and the distance covered by the car per hour is the amount in the fractional part $1\frac{1}{2}$ so we multiply the amount in 1 with the fractional part to find the amount in the fractional part:

$56 \times 1\frac{1}{2} = 84$ miles per hour

The car covers 84 miles per hour.

31. Wanda is reading a book of 320 pages. If she read $\frac{1}{8}$ of the book on the first day and $\frac{1}{2}$ as much on the second day as what she read on the first day, how many pages did she read in both days?

Reasoning and solution:

We need two sets of the 3 elements to solve the problem:

The first set of 3 elements:

From "she read $\frac{1}{8}$ of the book on the first day" we know that amount in 1 is the entire book, 320 pages and that the amount in the fractional part, $\frac{1}{8}$, is the number of pages she read on the first day. We multiply the fractional part with the amount in 1 to find the amount in the fractional part:

$320 \times \frac{1}{8}$ =40 pages

The second set of 3 elements:

From "$\frac{1}{2}$ as much on the second day as what she read on the first day" we know that the number of pages she read on the first day is the amount in 1 and that the number of pages she read on the second day is the amount in the fractional part, $\frac{1}{2}$, so we multiply the amount in 1 with the fractional part to find the number of pages she read on the second day:

$40 \times \frac{1}{2}$ =20 pages

The total number of pages she read in both days is:

40+20=60 pages

32. A clothing factory planned to make 720 children clothes in March but actually made $\frac{3}{5}$ of the amount during the first half of the month and the same amount in the second half of the month as in the first half of the month. How many more clothes did the factory actually make

in March?

Reasoning and solution:

We need only one set of the 3 elements to solve the problem:

From "actually made $\frac{3}{5}$ of the amount during the first half of the

month" we know that 720 clothes is the amount in 1 and that the number of clothes the factory actually made during the first half of the month

is the amount in the fractional part, $\frac{3}{5}$ so we multiply the amount in 1

with the fractional part to find the number of clothes the factory actually made during the first half of the month:

$$720 \times \frac{3}{5} = 432$$

Since the factory made "the same amount in the second half of the month as in the first half of the month" the total number of clothes made during the month is:

432+432=864

864−720=144 clothes

The factory actually made 144 more clothes than the amount they planned to make.

33. One class has 55 students in it; $\frac{1}{5}$ of them are girl students. When

5 boys are transferred to other classes, the number of girls is what percentage of the students in the class?

Reasoning and solution:

We need two sets of the 3 elements to solve the problem:

From "$\frac{1}{5}$ of them are girl students" we know that 55 students is the amount in 1 and that the number of girls is the amount in the fractional part, $\frac{1}{5}$, so we multiply the amount in 1 with the fractional part to find the amount in the fractional part, the number of girls:

$55 \times \frac{1}{5} = 11$ girls

When 5 boys are transferred to other classes, there will be 55−5=50 students in the class.

The second set of the 3 elements:

From "the number of girls is what percentage of the students in the class" we know that the number of students in the class, 50, is the amount in 1 and that the number of girls is the amount in a fractional part, so we divide the amount in 1 into the amount in the fractional part to find the fractional part:

$11 \div 50 = \frac{11}{50} = 22\%$

34. When the length and the width of a rectangle are both shortened to $\frac{2}{3}$ of the original size, the area of new rectangle is what fractional part of the area of the original rectangle?

Reasoning and solution:

Let L represent the original length and W represent the original width so the original area is LW.

When the length is shortened to $\frac{2}{3}$ of it, the shortened length should be

$\frac{2}{3}$ L; when the width is shortened to $\frac{2}{3}$ of it, the shortened width should

be $\frac{2}{3}$ W so the area of the new rectangle is:

$$\frac{2}{3} L \times \frac{2}{3} W = \frac{4}{9} LW$$

From "the area of new rectangle is what fractional part of the area of the original rectangle" we know that the amount in 1 is the area of the original rectangle and that the area of the new rectangle is he amount in a fraction so we divide the amount in 1 into the amount in the fraction to find the fraction:

$$\frac{4}{9} LW \div LW = \frac{4}{9}$$

35. The base of a parallelogram is increased to $\frac{7}{2}$ as long as the

original base; at the same time, the height on the base is reduced to

$\frac{2}{7}$ as long as the original height. The area of the new parallelogram is

what percentage of the area of the original parallelogram?

Reasoning and solution:

Let b represent the original base; let h represent the original height so the original area of the parallelogram is:bh

When the base is increased to "$\frac{7}{2}$ as long as the original base" the new base is $\frac{7}{2}$ b.

Then the height is reduced to "$\frac{2}{7}$ as long as the original height" the new height is $\frac{2}{7}$ h so the area of the new parallelogram is $\frac{7}{2}$ b$\times\frac{2}{7}$ h=bh

The area of the new parallelogram equals to that of the original one so the area of the new parallelogram is 100% of the area of the original area.

36. A factory planned to make 480 washing machines last week and $\frac{1}{6}$ more this week. If number of washing machines the factory plans to make next week is $\frac{1}{8}$ more than this week, how many washing machines does the factory plan to make next week?

Reasoning and solution:

We need two sets of the 3 elements to solve the problem:

The first set of the 3 elements:

From "$\frac{1}{6}$ more this week (than last week)" we know that the amount in 1 is the number of washing machines the factory planned to make last week and that the number of washing machines the factory planned to make this week was in the fractional part that is $\frac{1}{6}$ more than 1, which

is $1+\frac{1}{6}=\frac{7}{6}$ we multiple the amount in 1 with the fractional part to find the amount in the fractional part, the number of washing machines the factory planned to make this week:

$$480 \times \frac{7}{6} = 560$$

The second set of the 3 elements:

From "number of washing machines the factory plans to make next week is $\frac{1}{8}$ more than this week" we know that the number of washing machines the factory planned to make this week is the amount in 1 and that the number of washing machines the factory planned to make next week is the amount in the fractional part that is $\frac{1}{8}$ more than 1, which is $1 + \frac{1}{8} = \frac{9}{8}$ so we multiply the amount in 1 with the fractional part to find the number of washing machines the factory planned to make next week:

$$560 \times \frac{9}{8} = 630$$

37. There were 96 students in a math club and the number of boy students was $\frac{3}{8}$ of the number of all the students in the club. Recently, more boy students have joined the club so that the number of boys is $\frac{5}{6}$ of the number of girls in the club now. How many boy students have joined the club recently?

Reasoning and solution:

The first set of the 3 elements:

From "$\frac{3}{8}$ of the number of all the students in the club" we know that amount in 1 is the total number of students, 96 and that the number of boy students is the amount in the fractional part, $\frac{3}{8}$, so we multiply the fractional part with the amount in 1 to find the number of boys:

$96 \times \frac{3}{8}$ =36 boys

The number of girls is: 96−36=60

The 2nd set of the 3 elements:

From "so that the number of boys is $\frac{5}{6}$ of the number of girls now" we know that the number of girls is the amount in 1 and the number of boys is the amount in the fractional part, $\frac{5}{6}$ so we multiply the fractional part with the amount in 1 to find the number of boys now:

$60 \times \frac{5}{6}$ =50

The number of boys that joined the club: 50−36=14

38. Three groups of city workers are planting trees. The second group has already planted 480 trees; the third group planted $1\frac{1}{6}$ as many as the second group; the first group planted $\frac{5}{6}$ as many as the second group. How many trees have all the workers planted?

Reasoning and solution:

From "the third group planted $1\frac{1}{6}$ as many as the second group" we know that the number of trees planted by the second group is the amount in 1

56

and that the number of trees planted by the third group is the amount in the fraction $1\frac{1}{6}$ so we multiply the fraction with the amount in 1 to find the number of trees planted by the third group:

$480 \times 1\frac{1}{6}$ =560 trees

From "the first group planted $\frac{5}{6}$ as many as the second group" we know that the number of trees planted by the first group is the amount in 1 and that the number of trees planted by the first group is the amount in the fractional part, $\frac{5}{6}$ so we multiply the fractional part with the amount in 1 to find the amount in the fractional part, the number of trees planted by the first group:

$480 \times \frac{5}{6}$ =400 trees

The total number of trees planted by all the three groups is:

480+400+560=1,440 trees

39. A rope is 12 feet long. If $\frac{1}{5}$ of it is cut off at first and $\frac{1}{3}$ of it is cut off after that, how much shorter is the first cut than the second one?

Reasoning and solution:

Let's first find $\frac{1}{5}$ of the rope is how much less than $\frac{1}{3}$ of the rope:

$$\frac{1}{3} - \frac{1}{5} = \frac{2}{15}$$

We need to find the amount in $\frac{2}{15}$ of the entire rope:

$$12 \times \frac{2}{15} = \frac{8}{5} \text{ feet}$$

The first cut is $\frac{8}{5}$ feet shorter than the second one.

40. There are 3 piles of coal: the first pile has 240 tons of coal in it; the second pile of coal weighs $\frac{3}{4}$ as much as the first pile; the second pile of coal weighs $\frac{4}{5}$ ton less than the third pile of coal. How much do all the three piles of coal weigh?

Reasoning and solution:

From "the second pile of coal weighs $\frac{3}{4}$ as much as the first pile" we know that weight of the coal in the first pile is the amount in 1 and that the weight of the coal in the second pile is the amount in the fractional part, $\frac{3}{4}$ so we multiply the amount in 1 with the fractional part to find the weight of the coal in the second pile:

$$240 \times \frac{3}{4} = 180 \text{ tons}$$

From "the second pile of coal weighs $\frac{4}{5}$ ton less than the third pile of coal" we know that the weight of the coal in the third pile is:

$180 + \dfrac{4}{5} = 180\dfrac{4}{5}$ tons

The total weight of all the coal in all the 3 piles is:

$240 + 180 + 180\dfrac{4}{5} = 600\dfrac{4}{5}$ tons

41. Hanna is reading a book of 320 pages and has read $\dfrac{2}{5}$ of the book on the first day. If she read $\dfrac{3}{8}$ of the remaining pages of the book on the second day, how many pages hasn't she read yet?

Reasoning and solution:

From "read $\dfrac{2}{5}$ of the book on the first day" we know that the entire book is the amount in 1 and that the number of pages she read on the first day is the amount in the fractional part, $\dfrac{2}{5}$ so we multiply the fractional part with the amount in 1 to find the amount in the fractional part, the number of pages she read on the first day:

$320 \times \dfrac{2}{5} = 128$ pages

The number of pages she has not read yet:

$320 - 128 = 192$ pages

From "she read $\dfrac{3}{8}$ of the remaining pages of the book on the second day" we know that the number of pages she has not read is the amount in 1 and that the number of pages she read on the second day is the amount in the fractional part, $\frac{3}{8}$ so number of pages she read on the second day is:

$$192 \times \frac{3}{8} = 72 \text{ pages}$$

The number of pages Hanna has not read yet:

192−72=120 pages

42. A pile of coal weighs 40 tons; a truck shipped $\frac{2}{5}$ of it away on the first trip. Based on the information, answer the following 3 questions:

(1). If the truck ships away $\frac{3}{4}$ ton on the second trip, how many tons will be left after that?

Reasoning and solution:

From "a truck shipped $\frac{2}{5}$ of it away on the first trip" we know that the total amount of coal, 40 tons, is the amount in 1 and that the amount shipped away on the first trip is the amount in the fractional part, $\frac{2}{5}$ so the amount shipped away on the first trip is:

$$40 \times \frac{2}{5} = 16 \text{ tons}$$

The total amount of coal left after the second trip is:

$$40-16-\frac{3}{4} = 24-0.75 = 23.25 \text{ tons}$$

(2). If the truck ships away $\frac{3}{4}$ of the rest of the coal on the second trip, how many tons of coal will be left after that?

Reasoning and solution:

From (1) we know that the amount of coal shipped away on the first trip is 16 tons and rest of the coal after that should be 40-16=24 tons

From "the truck ships away $\frac{3}{4}$ of the rest of the coal on the second trip" we know that the rest of the coal is the amount in 1 and that the amount shipped away on the second trip is the amount in the fractional part, $\frac{3}{4}$ so the amount of coal shipped away on the second trip is:

$24 \times \frac{3}{4} = 18$ tons

The amount of coal left after the two trips was:

40-16-18=6 tons

(3). If the truck ships away $\frac{3}{4}$ as much as what it shipped away on the first trip, how many tons of coal will be left after that?

Reasoning and solution:

From (1), we know that 16 tons of coal was shipped away on the first trip. From "the truck ships away $\frac{3}{4}$ as much as what it shipped away on the first trip" we know that the amount shipped away on the first trip is the amount in 1 and that the amount shipped away on the second trip is the amount in the fractional part, $\frac{3}{4}$ so the amount shipped away on the second trip is:

$16 \times \frac{3}{4} = 12$ tons

The amount of coal left after the two trips is:

40−16−12=12 tons

43. A bookstore sold 1,400 books on the morning of a Saturday and $\frac{6}{7}$ as many on the afternoon of the Saturday. If the number of books sold on Sunday was $\frac{9}{13}$ as many as the number of books sold on Saturday, how many books were sold on Sunday?

Reasoning and solution:

From "$\frac{6}{7}$ as many on the afternoon of the Saturday (as on the morning of the Saturday)" we know that the number of books sold in the morning is the amount in 1 and that the number of books sold in the afternoon is the amount in the fraction $\frac{6}{7}$ so number of books sold in the afternoon is :

$1,400 \times \frac{6}{7} = 1,200$ books

The total number of books sold on Saturday is:

1,400+1,200=2,600

From "the number of books sold on Sunday was $\frac{9}{13}$ as many as the number of books sold on Saturday" the number of books sold on Saturday is the amount in 1 and the number of books sold on Sunday is the amount in the fraction $\frac{9}{13}$ so the number of books sold on Sunday is:

$$2,600 \times \frac{9}{13} = 1,800$$

44. The amount of money Peter has is $\frac{1}{5}$ more than John; the amount of money John has is $\frac{1}{5}$ more than Leon. The amount of money that Peter has is what fractional part more than the amount of money that Leon has?

Reasoning and solution:

Both Peter and Leon are related with John and we need to find the relationship between the amount of money Leon has and the amount of money that Peter has. Therefore, we will have 2 sets of 3 elements that are dependent on each other:

First set of the 3 elements:

From "the amount of money John has is $\frac{1}{5}$ more than Leon" we know that the amount of money that Leon has is the amount in 1 and that the amount of money that John has is the amount in the fraction that is $\frac{1}{5}$ more than 1 so the amount of money that John has is in the fraction:

$$1 + \frac{1}{5} = \frac{6}{5}$$

So the amount of money John has is $\frac{6}{5}$ of the amount of money that Leon has.

From "The amount of money Peter has is $\frac{1}{5}$ more than John" we know that the amount of money that John has is the amount in 1 but we need to compare the amount of money Peter has with the amount that Leon has though John's money. Therefore, we must keep the fraction that the John's money is in CONSTANT in all both sets of the 3 elements in the entire problem. So the amount of money that John has is in $\frac{6}{5}$ and the amount of money that Peter has is in the fraction that is $\frac{1}{5}$ more than $\frac{6}{5}$, which is $\frac{6}{5} + \frac{1}{5} = \frac{7}{5}$.

Remember: the amount in 1 is still the amount of money that Leon has.

So the amount of money that Peter has is $\frac{7}{5}$ of Leon's money.

To find the amount of money that Peter has is what fractional part more than the amount of money that Leon has we do the following:

We know that the amount of money that Peter has is $\frac{7}{5}$ of Leon's money and that the amount of money that Leon has is still in 1:

$$\frac{7}{5} - 1 = \frac{1}{5}$$

The amount of money that Peter has $\frac{1}{5}$ more than the amount that Leon has.

45. A book has 135 pages. Lily has read $\frac{2}{5}$ of the book during the first 5 days after she bought the book so which page will she need to start with on the 6th day?

Reasoning and solution:

From "has read $\frac{2}{5}$ of the book during the first 5 days" we know that the total number of pages is the amount in 1 and that the number of pages she read in these 5 days is the amount in the fraction $\frac{2}{5}$ so the number of pages she read in the 5 days is:

$135 \times \frac{2}{5}$ =54 pages

She needs to start with Page 55 on the 6th day.

46. A farmer harvested 6,400 pounds of apples last year. If this year the harvest was $\frac{1}{5}$ less than last year, how many fewer pounds of apples did the farmer harvest this year than last year?

Reasoning and solution:

From "this year the harvest was $\frac{1}{5}$ less than last year" we know that the weight of the apples harvested last year is the amount in 1 and that the weight of the apples harvested this year is the amount in $\frac{1}{5}$ less than 1, which is ($1-\frac{1}{5}$), or $\frac{4}{5}$ so the weight of the apples harvested this year is:

$6,400 \times \dfrac{4}{5} = 5,120$ pounds

$6400 - 5120 = 1,280$ pounds less

47. A store got a total of 960 pounds of apples and pears to sell. If the apples weighed $\dfrac{3}{5}$ of the fruits, how many more pounds of apples did the store get than pears?

Reasoning and solution:

Since the apples weighed $\dfrac{3}{5}$ of all the fruits and the store got only apples and pears to sell, the weight of the pears should be $(1 - \dfrac{3}{5})$ or $\dfrac{2}{5}$ of all the fruits.

How much more apples did the store gets than pears? Here is the answer:

$\dfrac{3}{5} - \dfrac{2}{5} = \dfrac{1}{5}$

$\dfrac{1}{5}$ of all the fruits are:

$960 \times \dfrac{1}{5} = 192$ pounds more apples than pears

48. A construction crew is maintaining a road and has already worked on it for 5 days. If, on the average, the crew fixes $\dfrac{1}{20}$ of the road on each day, what fractional part of the road hasn't the crew maintained yet?

Reasoning and solution:

From "the crew fixes $\frac{1}{20}$ of the road on each day" we know that the entire road is 1 and that on each day the crew fixed $\frac{1}{20}$ of it. The crew worked on the road for 5 days so the fraction of the entire road the crew fixed is:

$$\frac{1}{20} \times 5 = \frac{1}{4}$$

Deduct the part of the road fixed from the entire road to find what part of the road is not fixed yet:

$$1 - \frac{1}{4} = \frac{3}{4}$$

49. The distance between City A and City B is 150 miles. Peter started from City A and is driving to City B. If he has travelled $\frac{2}{5}$ of the entire distance, how far is he away from City B?

Reasoning and solution:

He has covered $\frac{2}{5}$ of the entire distance so the entire distance is 1 and the distance he has not covered is $1 - \frac{2}{5} = \frac{3}{5}$

$\frac{3}{5}$ of the entire distance is:

$$\frac{3}{5} \times 150 = 90 \text{ miles}$$

Peter is still 90 miles away from City B.

50. A rope is $3\frac{2}{5}$ feet long. Peter cut $\frac{1}{2}$ of it at first then cut $\frac{1}{2}$ foot from the rest of the rope. How many feet of the rope are still left?

Reasoning and solution:

$\frac{1}{2}$ of the rope is the amount in the fraction $\frac{1}{2}$ and the total length of the rope is the amount in 1 so $\frac{1}{2}$ of the rope is:

$$3\frac{2}{5} \times \frac{1}{2} = \frac{17}{10} \text{ feet}$$

Deduct $\frac{1}{2}$ foot from the rest of the rope:

$$\frac{17}{10} - \frac{1}{2} = 1\frac{1}{5} \text{ feet left}$$